Hugh Rodman

Report of Ice and Ice Movements in the North Atlantic Ocean

Hugh Rodman

Report of Ice and Ice Movements in the North Atlantic Ocean

ISBN/EAN: 9783337038052

Printed in Europe, USA, Canada, Australia, Japan

Cover: Foto ©berggeist007 / pixelio.de

More available books at **www.hansebooks.com**

REPORT

OF

ICE AND ICE MOVEMENTS

IN THE

NORTH ATLANTIC OCEAN,

BY

ENSIGN HUGH RODMAN, U. S. N.

UNDER THE DIRECTION OF

CAPT. HENRY F. PICKING, U. S. N.
HYDROGRAPHER.

WASHINGTON:
GOVERNMENT PRINTING OFFICE.
1890.

INTRODUCTION.

The subject of ice and ice movements is one to which the Hydrographic Office has given special attention, mainly through its branches at Boston and New York, where reports are regularly received and transmitted to the meteorological division of this Office; here the information is carefully tabulated, and appears on the monthly pilot-chart of the North Atlantic Ocean, on which the ice for the previous month is plotted, and a course laid down to clear it. This course, though some 200 miles longer than the shortest possible route (close to Cape Race), is the most ecomical one that can be followed, not only clearing the ice region, but also avoiding that greatest menace to navigation, fog.

At first these precautionary measures were deemed excessive by many captains, but the number is notably decreasing annually, and their wisdom is becoming apparent in the gradual approximation to the pilot-chart routes by the trans-Atlantic steamer lines.

As the season advances, the force of the Boston and New York branch offices is greatly occupied in furnishing information on this important subject to captains, for whom, on the eve of sailing, all the latest ice reports are plotted on the charts, and the managers of the steam-ship lines are supplied with copies of all ice reports received.

The opening of navigation in the St. Lawrence River and Gulf is an important matter in the New York market, and the subsequent movement of the ice is a subject of great interest to those wishing to charter vessels for that trade.

The Office has made every effort to collect the latest and most reliable data concerning the movement of ice in the vicinity of the Grand Banks, and the officer in charge of the Boston branch, in a previous report, states that the marked decrease in the supply of information on the subject was due to the fact of the steamer lines adopting the routes recommended by this Office.

It is hardly necessary to dwell upon the importance of the ice problem, or the benefit to be derived from a thorough understanding of it; and, in order that the maritime community might better appreciate the danger, and be aided in intelligently studying the best means of avoiding it, at the request of the Hydrographer, Ensign Hugh Rodman, U. S. Navy of the division of marine meteorology of this Office, was detailed by the bureau, and orders issued by the honorable Secretary of the Navy to proceed to Halifax, Nova Scotia, and St. John's, Newfoundland, and make a careful and thorough report on the subject, collecting the most reliable and authentic data, and enlisting the most extensive co-operation. Special ice forms were prepared and distributed to the Newfoundland sealers, who are the first to sight the heavy Arctic field ice; to the whalers of the northern fleet; to a number of stationary observers in Newfoundland and Labrador, and to all, in fact, who sight the ice or spend the season in the ice regions, from which, no doubt, the greatest benefit in the future will be derived, as it goes without saying that the greater the number of reports received the more accurate the resulting predictions.

The accompanying report by Ensign Rodman refers more particularly to ice as an obstruction to transatlantic navigation, and for this reason many other interesting points are merely touched upon, others are being more carefully elaborated, and an effort is making to trace the ice-flow from beginning to end.

The appreciation of the maritime community for the report of ice movements in Bering Sea and the Arctic basin (U. S. Hydrographic Office publication No. 92) led the Hydrographer to endeavor to issue a similar report for the North Atlantic Ocean, the outcome of which was the accompanying report.

This Office is especially indebted to Capt. William Maxwell, R. N., Halifax, Nova Scotia, for his courtesy to its representatives in furnishing most of the data in the appended forms relative to the opening and closing of ports and for other valuable suggestions and assistance; also to Mr. J. T. Nevill, inspector of light-houses, St. John's, Newfoundland, through whose kindness the keepers will, in future, keep a regular form, which will be forwarded to us as soon as filled in order that our published information may be as complete as possible.

HENRY F. PICKING,
Captain, U. S. Navy, Hydrographer.

U. S. HYDROGRAPHIC OFFICE,
Washington, May, 1890.

ICE AND ICE MOVEMENTS IN THE NORTH ATLANTIC.

In the investigation of ice as an obstruction to navigation in other than Arctic waters it will only be necessary to dwell briefly upon its formation in that region, and to follow it in its course out of the Arctic basin to the southward until it finally disappears off the Banks and along the course of the Gulf Stream. · Not that all ice-fields have their origin there, for the coasts of Newfoundland and Labrador and the Gulf of St. Lawrence are prolific in their supplies, and send out yearly miles of ice-fields, which cause much delay and danger, block ports, and interfere seriously with the deep-sea and coasting trade. Many names are given to different kinds of ice, often the same kind is called by different names, and to the uninitiated this would be confusing, so only the ordinary terms will be used, which explain themselves. In order to understand the ice-drift a knowledge of currents, winds, and tides are very necessary as well as a general idea of the contour of the coasts along its paths.

In deep water the currents can be easily traced and plotted on charts, but as the shore line is approached they often become very erratic, and a local knowledge is necessary before these peculiarities can be fully understood.

No ocean current should be considered as always being within the same limit throughout its entire length, like a river inclosed within its banks, like the lower Hudson for instance, which never overflows its banks, but a variation must be allowed to all of them. The limits of the Gulf Stream can be defined and remain about the same each year and month between Florida and Cuba; its strength may vary at times even here but its limits do not, while north of Cape Hatteras its northern and southern limits vary considerably, both annually and after severe gales of wind, especially those blowing at a great angle to its axis. In the same way the Arctic current must vary in its limits.

There are really two currents which transport ice, the Arctic and East Greenland; the latter, flowing southwesterly along the East Greenland coast, rounds Cape Farewell, flows northwesterly to about latitude 63° in Davis Strait, turns upon itself and unites with the regular current flowing south through Davis Strait; thence it sweeps the coast of Labrador, spreading to the eastward in its course until about latitude 52° is reached, when the eastern edge turns towards the eastward, a small part finds its way through the Straits of Belle Isle into the Gulf of St. Lawrence, while the main body of the stream, continuing to the southward, rounds Newfoundland and forms an inshore current as far south as Florida.

It will be noticed that off the Banks the axis of the Gulf Stream and that of the Arctic current are nearly at right angles, which will cause a maximum variation in the northern limit of the Gulf Stream depending upon the force of the Arctic current as it may be augmented or diminished by the wind's influence, and this variation will be greatest when the Gulf Stream is at its annual southern limit. The line of demarkation must at all times be formed by the resultant of the two forces depending upon the relative strengths of the currents.

Many suppose that near the line of meeting the Arctic current sinks under the Gulf Stream, so that the northern surface current of the Gulf Stream flows easterly over a lower cold current setting to the southward and westward. This would appear to be true from the path of many bergs, and will be analyzed more fully when the drift of the bergs is considered.

Along the entire coasts of Labrador and Newfoundland and in the Gulf of St. Lawrence the shore currents are very confusing and variable. Add to this tidal influences and the force and direction of the wind, and the problem becomes one which can only be solved by an extensive local experience. Even then there are often such variations that all knowledge fails, and hundreds of vessels have been lost through lack of knowledge on the part of the best local pilots. It is unnecessary to go into detail in this work of the currents in each locality, but rather to give a general idea which can apply to the subject as a whole.

BERGS.

The bergs which annually appear in the North Atlantic have their origin almost exclusively in West Greenland. Indeed, this is the great berg factory, and although a few may round Cape Farewell from the Spitzbergen Sea, and a few come from Frobisher Sound and Hudson Strait, we may still give Greenland the credit of furnishing at least eighty per cent. of all that are made that come to the southward. The interior of Greenland is a solid mass of ice, covered by snow, while a narrow belt around the coast is the only part uncovered. Extending from the interior through this belt are the numerous ice fjords or glaciers which, including large and small, number in the hundreds. These range in width from a few hundred feet to several miles, and from 50 to 1,500 feet in thickness. All of these glaciers are making their way towards the sea, and as their ends are forced out into the water from the pressure behind they are broken off and set adrift as bergs. The rate of movement of the glaciers is variously estimated, and of course varies in the different ones and at different seasons, but from observations made they have been known to advance at the rate of 47 feet a day. Some observers have attempted to connect this rate with the amount of water discharged by the subglacial stream, but in Greenland it has not proved satisfactory.

Once the glacier extends into deep water pieces are broken off by their buoyancy, aided possibly by currents and by the brittleness of the ice. This process is called *calving;* the size of the pieces set adrift varies greatly, but a berg from 60 to 100 feet to the top of its walls, whose spires or pinnacles may reach from 200 to 250 feet in height and from 300 to 500 yards in length, is considered an average size berg in the Arctic. These measurements apply to the part above water, which is about one-eighth or one-ninth of the whole mass. Many authors give the depth under water as being from eight to nine times the height above; this is incorrect, and measurements above and below water should be referred to mass and not to height. It is even possible to have a berg as high out of water as it is deep below the surface, for if we imagine a large, solid lump, of any regular shape, which has a very small sharp high pinnacle in the center, the height above water can easily be equal to the depth below. An authentic case on record is that of a berg grounded in the Strait of Belle Isle in 16 fathoms of water that had a thin spire about 100 feet in height.

Many estimates have been made of the number of pieces, or bergs, that each glacier puts out on an average each year, and it is given at from 10 to 100, while the mass might average 250,000,000,000 cubic feet.

From observations made on a particular glacier the following figures will show its output: Breadth, 18,400 feet; depth, 940 feet; advance per day, 47 feet during summer season. This would give about 200,000,000,000 cubic feet a year, as the product of an average size ice fjord, which, allowing 5 pounds a day for each person in the United States, would last over one hundred years.

Bergs are made all the year round, but the greater number during the summer season, so that thousands are set adrift each year.

Once adrift in the Arctic they find their way into the Arctic current and begin their journey to the southward. It is not an unobstructed drift, but one attended with many stoppages and mishaps. Many ground in the Arctic basin and break up there; others reach the shores of Labrador, where from one end to the other they continually ground and float; some break up and disappear entirely, while others get safely past and reach the Grand Banks. A glance at the chart will show how the set of the Arctic current is on the shore of Labrador, which shunts it finally a little to the eastward. This whole coast is cut up by numerous islands, bays and headlands, shoals and reefs, which makes the journey of all drift a long one, and adds greatly to the destruction of the bergs by stoppages and causing them to break up.

It must not be supposed that all bergs made in any one season find their way south during the following one, for only a small percentage of them ever reach transatlantic routes. So many delays attend their journey, and so irregular and erratic is it, that many bergs seen in any one season may have been made several seasons before. If bergs on their calving at once drifted to the southward, and met with no obstructions, their journey of about 1,200 to 1,500 miles would occupy from 4 to 5 months, reckoning the drift of the Arctic current at 10 miles a day, which may be making it too little. Then if bergs were liberated principally in July and August they should reach transatlantic routes in December and January, while we know this to be the rare exception. It is then seen what an important bearing the shores of Labrador have in arresting their flow, when it is known that bergs are generally most plentiful in the late spring and early summer months off the Banks.

It should not be supposed that all bergs follow the same course when set adrift from their parent glaciers, for, like floating bodies at the head of a river, some will go direct to the mouth, others will go but a short distance and lodge, others still will accomplish half the journey and remain until another freshet again floats them, so that in the end the débris will be composed in part of that of several years' production.

Bergs, when first liberated on the west Greenland shore, are out of the strongest sweep of the southerly current, and they may take some months to find their way out of Davis Strait, while again another may at once drift into the current and move unobstructed until demolished in the Gulf Stream. The difference in time of two bergs in reaching a low latitude which were set adrift the same day may cover a period of one or two years.

Field ice also offers an obstruction to bergs, and a close season in the Arctic may prevent their liberation to a great extent, though, from their deep submersion, they act as ice-plows and aid materially in breaking up the vast fields of ice which so often close the Arctic Basin.

Ice-fields are more affected by wind than bergs, as will be shown hereafter, while bergs owe their drift almost entirely to current, so that they will often be noticed forcing their way through immense fields of heavy ice and going directly to windward. Advantage is taken of this by vessels in ice-fields, which often moor to bergs and are towed for miles through ice through which they could not otherwise make any headway. This is accomplished by sinking an ice-anchor into them and using a strong tow line, and as the berg advances open water is left to leeward while the loose ice floats past on both sides. For the same reason vessels, when beset by field ice, run from the lee of one berg to that of another, as leads may offer themselves.

All ice is brittle, especially that in bergs, and it is wonderful how little it takes to accomplish their destruction. A blow of an ax will at times split them, and the report of a gun, by concussion, will accomplish the same end. They are more apt to break up in warm weather than cold, and whalers and sealers note this before landing on them, when an anchor is to be planted or fresh water to be obtained. On the coast of Labrador in July and August, when it is packed with bergs, the noise of rupture is often deafening, and those experienced in ice give them a wide berth.

When they are frozen the temperature is very low, so that when their surface is exposed to a thawing temperature the tension of the exterior and interior is very different, making them not unlike a Prince Rupert's drop. Then, too, during the day water made by melting finds its way into the crevices, freezes, and hence expands, and, acting like a wedge, forces the berg into fragments.

The effect of a change in temperature of the outside can be readily shown by the following experiment:

Take a large block of ice, frozen at a temperature of 5° F., and suddenly immerse it in water above the temperature of 90° F., and the block will immediately go to pieces, making a noise that can be heard several yards distant. This shows the ice is not only destroyed by slow melting from the outside but by rupture, thus bringing a greater surface in contact with warmer water, which promotes melting. This is what happens to bergs when exposed to a warmer atmosphere, or when brought in contact with the Gulf Stream. If a large berg would remain intact and not go to pieces, several years might elapse before it would melt, and they would be ever present in transatlantic routes, and might at times reach well over towards the European coast.

By some it is supposed that bergs contain organic substances which are frozen in them, which promote melting by the radiation of the sun's heat, but this is not clearly demonstrated.

They assume the greatest variety of shapes, from those approximating to some regular geometric figure to others crowned with spires, domes, minarets, and peaks, while others still are pierced by deep indentations or caves. Small cataracts precipitate themselves from the large bergs, while from many icicles hang in clusters from every projecting ledge. They frequently have outlying spurs under water, which are as dangerous as any other sunken reefs. For this reason it is advisable for vessels to give them a wide berth, for there are a number of cases on record where vessels were seriously damaged by striking when apparently clear of the berg. Among these is that of the British steamship *Nessmore*, which ran into a berg in latitude 41° 50' N., longitude 52° W., and stove in her bows. On docking her a long score was found extending from abreast her fore-rigging all of the way aft, just above her keel. Four frames were broken and the plates were almost cut through. The ship evidently struck a projecting spur after her helm had been put over, as there was clear water between her and the berg after the first collision.

It is best to go to windward of them, for débris broken off will generally drift to leeward and open water be found to windward.

There are examples without end where vessels have been seriously injured, if not lost, by bergs breaking up or turning over. Often they are so nicely balanced that the slightest melting of their surfaces causes a change in their center of gravity, when they are liable to capsize. In the same way if bergs nearly in the state of unstable equilibrium when drifting with the current and their bottom grounds will frequently turn over and break to pieces.

Many bergs, after leaving the coast of Labrador, where they are generally much reduced in size, find their way to the Newfoundland coast and ground there and on the edge of the Grand Banks, where their destruction continues.

There are more ways than one of telling one's proximity to bergs, which, fortunately for vessels, make their presence known in all kinds of weather.

On a clear day they can be seen at a long distance, owing to their brightness, and at night to their effulgence. During foggy weather they are seen through the fog by their apparent blackness, if such a term can be applied, though it is never safe to run at full speed in thick weather when in the vicinity of bergs, as many vessels have found to their cost.

They can be detected by the echo from the whistle or fog-horn, and this point should be remembered and noted. By noting time between blast of whistle and return sound, the distance of the object can be found approximately, in feet, by multiplying by 550.

The presence of bergs is often made known by the noise of their breaking up and falling to pieces. Temperature is useful in indicating the presence of ice, though near the junction of the Arctic current and Gulf Stream great variations will always be found. If a berg be grounded, or if the surface current be the stronger, water flowing past ice will be lowered in temperature and show its presence. To leeward of ice the atmospheric temperature is lowered, though like the temperature of water in ice regions it is subject to rapid changes. Change to a lower temperature, therefore, should serve as a warning, and a sharp lookout be kept, though it does not always follow that the change is caused by ice.

FIELD ICE.

Field ice is made from the Arctic to the shores of Newfoundland, and yearly leaves the shore to find its way into the path of commerce. Starting with the Arctic field ice and coming to the southward, we find this ice growing lighter, both in thickness and in quantity, until it disappears entirely. Ice made in the Arctic is heavier and has lived through a number of seasons. After the short summer in high latitudes ice begins to form on all open water, increasing several feet in thickness each season. Much of this remains north during the following summer, and though it melts to some extent it never entirely disappears, so that each succeeding winter adds to its thickness.

This continues from year to year until it reaches 12 or 15 feet in thickness, often more. If it remained perfectly quiet it would be of uniform thickness, increasing with the latitude, but it is in a state of almost continual motion, often a very violent one, which causes it to raft and pile until it becomes full of hummocks and other irregularities. Immense fields are detached from the shore and from other fields, and under the influence of winds, currents, and tides are set in motion and kept continually drifting from place to place; after a snow, thaw, or piling the whole becomes cemented together into solid pieces, when under the influence of a low temperature. The space of open water between the fields becomes frozen, joining smaller fields, and making a solid pack which will remain so until the elements again break it to pieces. Along the shores from headland to headland the bays and inlets often remain solid for years, almost invariably through the Arctic winter, but in Baffin Bay and Davis Strait open water can be found at intervals all the year round.

Ice becomes rafted in a variety of ways. If two fields are adrift the one to windward will drift down on the one to leeward; the one which is rougher on its surface gives the wind a better hold, and drifts the faster; fields may be impelled towards each other by winds from contrary directions. Ice that is secure to the shore is rafted on its seaward edge from contact with that which is adrift. Fields in drifting often have a turning motion, which is caused by contrary currents, or one variable in strength at different places, or by the friction of a field coming in contact with another field afloat or one attached to the shore. This rotary motion is especially dangerous when a vessel finds itself between two fields. A heavy gale will break up the strongest fields at times and cause them to raft and form hummocks.

Small fragments of bergs find themselves mingled with Arctic fields and become frozen fast. These, when liberated to the southward, are called *growlers*, and form low, dark, indigo-colored masses, which are just awash and rounded on top like a whale's back. They are very dangerous when in ice fields which have become loose enough to permit the passage of vessels through them, and should always be looked for; they can be seen apparently rising and sinking as the sea breaks over them.

During the spring and summer months the bergs, aided by a rise of temperature, so cut up and weaken the ice fields that much ice is loosened and begins drifting out of the Arctic Basin. This is joined by that brought from the Spitzbergen Sea by the East Greenland current, amalgamates in a fashion, whence it flows down the eastern coast of North America, reaching Cape Chidley about October or November. By this time the remaining ice in the Arctic is being cemented into solid fields, while the ice cap is being daily extended to the southward. As fast as fields are detached, the open water freezes, and these masses are forced to the southward and can not rejoin the solid pack. With a westerly wind ice formed in Hudson Strait and adjacent waters is swept out and joins the Arctic ice, differing from it only in being a little lighter.

Ice begins to form at Cape Chidley about the middle of October, at Belle Isle about November 1, and by the middle of November or 1st of December, the whole coast is solidly frozen. It should be stated here that the dates given are approximate and vary from year to year, marked exceptions being on record to many of them, so they represent averages.

The string of ice along the coast of Labrador extends from headland to headland, including the outlying islands, and starting from the heads of the bays, works its way out to seaward, forming by the middle of December an impassable barrier to the shore which will probably not be permanently broken until the latter part of April. This ice varies in thickness from 12 feet at the northern extreme to 3 or 4 feet at the southern. During the entire winter the Arctic drift is finding its way down the coast, and is being continually re-inforced by fields broken from the Labrador ice. These continue to the southward in the Arctic current on an average of about 10 miles a day, reaching Belle Isle between the middle of January and the middle of February.

The best example on record of a continued drift from the Arctic is that of Captain Tyson. On October 14, 1871, he and a party of nineteen others were separated from the United States surveying ship, *Polaris*, in latitude 77° or 78° N., just south of Littleton Island, and being unable to again reach the ship, remained on the floe and accomplished one of the most wonderful journeys on record. After a drift of over 1,500 miles, fraught with danger from beginning to end, not only from starvation and exposure to cold, but with death a number of times from the numerous gales encountered, supporting themselves by hunting and fishing, they were picked up about six months later, April 30, 1872, by the *Tigress*, a sealing steamer from Newfoundland, near the Strait of Belle Isle, in latitude 53° 35′ N., and carried safely into port. No better example could be given than this of the drift ice from the Arctic Basin; illustrating, as it does, not only the journey to the southward, but the many vicissitudes to which ice is subjected before reaching a low latitude.

Much delay will be caused by winds from the southward of west, as field ice is affected more by wind than current. The prevailing wind and weather will influence the drift very greatly. Strong northerly or northwest winds will increase its speed, contrary winds the reverse. The string of shore ice keeps the northern ice off the coast and in the current. At times westerly winds will also send the Labrador ice off the coast and leave it entirely clear, but this does not happen often. Still the outer Labrador ice is constantly being added to the Arctic flow. Frequently the bays remain frozen over until June, again, they are cleared some years in April, making a large variation. During the drift the wind from northwest to southwest will clear the ice off the coast and leave a line of open water, but the ice will be set on the coast by a northeast wind, and be rafted and piled. The appearance of the ice when it reaches Belle Isle and to the southward, would be a fair indication of the weather it had encountered on its way down, the rougher the ice the more severe the weather, and *vice versa*. This floating ice

string extends approximately 200 miles off shore, in the latitude of
Cape Harrison, and spreads more during its drift, though narrower
farther north. One small stream finds its way through the Strait of
Belle Isle, while the greater part continues to the northern limit of the
Gulf Stream. The shores of Newfoundland and Gulf of St. Lawrence
are full of ice, frozen there, by the middle of January, and are opened
or closed by a favorable or adverse wind. Navigation in the river St.
Lawrence is closed about the middle of November and does not open
until about May. A wind from northwest to southwest will clear the
eastern coast of Newfoundland, while the Gulf of St. Lawrence may
remain full of ice until the 1st of May. Even after this date much ice
is found in the Gulf until July, and by August or earlier the field ice is
replaced in the Strait of Belle Isle by bergs.

In the bight from Cape Bauld to Fogo Island a string of ice is often
found joining these points, hemming in the shore for weeks at a time.

With each northwest or westerly wind the ice is cleared off the Newfoundland coast, except from some of the deeper bays, and carried out
to sea, and frequently before the Arctic and Labrador ice has passed
Belle Isle the Newfoundland ice has found its way in the track of transatlantic vessels as far as 45° N. In the same way the Labrador ice
sometimes precedes the Arctic ice, while all may arrive at nearly the
same time. Ice fields often lose their identity, as coming from any one
particular place, by the constant intermingling on its southern journey
with ice made in a lower latitude.

Young slob ice may be found around the coast of Newfoundland from
December until April, some of which is heavy enough to prevent the
passage of vessels other than those specially built to go through it.

Ice leaving the Gulf of St. Lawrence flows south through Cabot
Strait, drifts southward with a few degrees of westing, its southern
limit depending upon its thickness, strength of current, or prevailing
wind, but often reaches Sable Island. This string of ice in 45° N. will
vary from a few miles to 50 in width. This can be easily avoided by
running to the southward and around it, or by shaping a course to clear
it between ports of departure and destination. The southern coast of
Newfoundland from Cape Ray to Cape Pine is generally clear of heavy
field ice, though the extreme eastern and western ports are occasionally
closed by it. Ice coming from the northward of Cape Race will round
it, spreading to the westward, sometimes filling Trepassy, St. Mary's,
and Placentia Bays, especially with a southerly wind.

There is no region affected by ice where its sudden appearance and
disappearance is not noticed. A harbor or locality may be entirely free
of ice—in a few hours a solid pack will be found; and, on the other hand,
a pack which has lasted for days will disappear in a few hours. All
ice seems to have a motion within itself, as well as one of drift as a
whole. Many instances are cited where two vessels have been jammed
in a field within a few cable lengths of each other, and at the end of

several days be a number of miles apart, yet there had never been a time when free communication on foot over the ice could not be maintained. In the same way vessels imprisoned in the ice often with a slight shift of wind find themselves freed with an open lead before them.

Ice fields assume a variety of shapes depending upon the influence of winds and currents, and upon their shape on being set adrift. Those loosened in the Arctic meet with so many vicissitudes that they have entirely lost their original form when a low latitude is reached, while those from Newfoundland may remain approximately intact. Their extent is governed by the same rules, and varies from a few scattered pieces to several hundred miles in length. Suppose, for example, a field of Arctic ice 20 miles in diameter to be nearly circular when detached in latitude 70° N. This may be composed of pans from a few feet to half a mile in diameter, 10 to 15 feet in thickness, with a fairly smooth surface. As it comes to the southward it encounters a series of gales from different points of the compass, it is brought in contact with the Labrador ice string, and finds itself in places where the current is of unequal strength on its opposite sides. It will then be lengthened or compressed, making a long irregular strip, possibly scattered, or be rafted and tighter. During its journey ice formed at other points gets mixed with it, so that by the time it reaches Belle Isle, its shape will be very irregular, long tongues extending from the mass, one side very tight, the opposite loose and scattering. Bergs may have plowed their way through it, or it may have been divided by an island, or by a gale, so that it may assume any form.

From off Belle Isle it finds its way south to the Gulf Stream, where no definite shape can be given it. In appearance, if heavy ice, it will be white, covered with snow, and visible at a long-distance; even in foggy weather it can often be seen for some distance when the fog is thickest. It is full of hummocks, and its surface is very uneven, blocks have been piled upon each other, others stood on end, and the whole mass will form an impenetrable field, through which vessels can not force their way.

If the ice be lighter the pans will be smoother and more even, the angles ground down by friction and turned up at the edges like so many large pond lilies. If compact no water is seen, if loose wide leads may extend through the whole, or a little water be seen surrounding each cake.

The appearance must decide whether a vessel is warranted in trying to force her way through. In a smooth sea, where doubt exists, should a vessel go dead slow into the mass, there will be but little danger in attempting it, and if too heavy, she can easily haul out. Often the weather edge is the heaviest from being rafted, when to leeward it may be scattering. An ice field will often form a good lee for riding out a gale of wind, as it will break the force of the sea. But care is necessary not to lie too close, for then the pans are often given such a force that they will stave in the bows of the strongest vessels.

The great difference then between Arctic ice and that made farther south is that the former takes several seasons to make, while the latter is made each season. Ice will form in open water in a cold calm night; in the latitude of St. John's, Newfoundland, it has formed 1 or 2 inches in thickness in the bays, and for 40 miles out to sea. Wherever it is protected for any length of time it will form several feet in thickness, though it is easily broken up by a heavy swell or strong wind. If the bay is open an off-shore wind soon empties it, but if full of islands it will take a much longer time.

Naturally the ice is looser to the southward than to the northward, so that the harbors on the east coast of Newfoundland are alternately being opened and closed during the entire ice season, while those farther north remain permanently closed for more than half the year. Hudson Strait, for example, is open to navigation for about three months, from July to October, but there is never a time when ice may not be encountered, and of such a heavy nature that a vessel can not force her way through it.

This does not mean that a vessel can only enter during these months for open water exists at times in almost every month, but for such short intervals that it would be useless to depend upon it.

As has been before stated, and from the foregoing remarks it will be seen that exact dates can not be given, and only approximate averages should be considered, and even these are often at fault from lack of data, and from conflicting reports which are given by those equally experienced in ice.

A table is appended showing a summary of observations wherever obtainable.

Many interesting points bearing on the ice formation and its destruction need only be referred to in this paper, as it is only intended to treat of ice as an obstruction to navigation. Much of the ice encountered at sea is discolored and often full of dirt and gravel, while not infrequently stones are found imbedded in it. Along the shores of Labrador, where there is a large rise and fall in the tide, ice is brought in contact with the bottom, and mud and seaweed are frozen in with it, while at times land slides precipitate large quantities of dirt and stones on its surface. As the ice leaves the coast and comes to the southward it brings these burdens with it, which are deposited on the ocean bottom when the ice melts. As this melting occurs to a great extent over the Grand Banks, it would seem that the deposit from the field ice would be greater than that from bergs. It is hard to understand why bergs should have foreign substances frozen into them, as they are formed from snow deposited on the frozen surfaces in the interior of Greenland, and hence their thickness is added to from their upper surface. It is possible that in their journey south in the Arctic current, they accumulate more or less foreign matter by having it ground into their bottoms, but this does not seem probable, as it is hard to force gravel into ice and give it a permanent hold, while mud accumulated in this way would soon be washed out. Then, too, the largest bergs find their way around the edges of the Banks and do not cross, on account of their draft, for only an average size berg crosses the Banks.

All field ice, no matter whether made near shore or at sea, when turned over shows a marked discoloration, which is called "foxy slime," yellowish red in color, which seems to rise from the bottom of the sea to the under surface of the ice. It is a well-known fact that the cod follow the Arctic ice down and feed under it, and the seals use it for breeding purposes.

It is of interest to know that all ice is not formed on the surface; it has been known to freeze in 10 or 15 fathoms of water on the Labrador coast. Seal-nets are set in these depths, and often the seals are found frozen solid at a depth of 12 fathoms, while huge pieces of ice form around the killick with which the nets are anchored and bring them to the surface. These killicks are made of stone, so the formation takes place around them, and when a sufficient quantity has been frozen the whole mass rises to the surface. A very marked example of bottom freezing is recorded where a box of iron tools, which was lost from a ship crushed in the ice in Hudson Strait, was afterwards found off Nain solidly frozen in a mass of ice, which had formed at the bottom around it and buoyed it to the surface, whence the current swept it to sea and landed it on the coast of Labrador, several hundred miles from where it was lost.

It is said of Labrador fishermen, who often buoy their seal-nets, that when a buoy is used the nets freeze and rise to the surface, but when not buoyed they seldom rise, indicating that the ice follows down the buoy-rope and reaches the net in this way. It would appear that ice frozen on the bottom would thus have a better chance of incorporating foreign bodies, and may be the greatest bearer of such burdens to the Grand Banks.

There can be no doubt that ice freezes on the bottom, for numerous instances are on record where vessels have observed it rise to the surface in open water, and a few instances where it has come up under a vessel, causing her to leak, or doing other damage.

Having now briefly outlined the drift of ocean ice from its source to its position in paths of commerce, let us see what its movements then are, and the causes which lead to its destruction.

From observations extending over a series of years it is shown that field ice reaches latitude 46° N. generally in February, varying from January to March, between the meridians of 46° and 50° W., while bergs reach the same locality from January to April. The marked exception has been the present season, when bergs came down in December, 1889, and field ice followed early in January. The drift of both bergs and field ice from this position during the month following the one in which they are first sighted will be about 200 miles south-southwest, the mass at the same time spreading along its southern limit. During the next month it will have reached the northern average limit of the Gulf Stream, and spread itself along this line, both east and west, when its destruction begins.

After reaching its southern limit many bergs, from their deep submersion, find their way to the westward, even when within the limits of the Gulf Stream, while field ice never follows this course.

This is accounted for by the fact that the Arctic current runs under the Gulf Stream, and a thin tongue of the latter lies over it. Each year, as this ice comes down, it is followed by other ice, making the season of field ice about three and a half months long, the average time of its disappearance being in April or early May, while the season of bergs lasts about seven months. The time of disappearance of bergs is generally from August until October 1. They have been reported as late as November, and, indeed, there is not a month in the year that they are not found in transatlantic routes in different years.

There is one locality in which it appears that ice of all kinds is most apt to be found during the regular season, say from March to June, and that is between latitude 42° and 45° N., and longitude 47° and 52° W., and this is probably due to the fact that the Gulf Stream and Arctic current meet here, and it is influenced sometimes by one, sometimes by the other. Should it drift bodily into the Gulf Stream it will be carried to the eastward, while it will drift to the westward should it remain in the Arctic current.

Many of the large bergs ground all around the edge of the Grand Banks, and those that round the extreme southern point in latitude 43° N., longitude 50° W., under the influence of the northerly current, will, when the limits of the currents change, as they continually do, be swept to the northward and eastward and ground to the westward of the Banks. Numerous reports show this to be a well-recognized fact.

There are wide variations in the times of appearance and disappearance of both bergs and field ice, as well as in their drift from year to year. Hardly any two seasons will be exactly similar but the above general rules will give averages.

During the year 1889 hardly any reports were made of ice until April, when numerous bergs, but hardly any field ice, were reported from latitude 48° to 46° N., between longitude 44° and 49° W. During May a large increase in the number of bergs follows, extending from 49° to 46° N., between 42° and 50° W. In June, from 49° to 43° N., between 45° and 50° W. In July, from 49° to 45° N., between 45° and 50° W. In August scarcely any was reported south of 50° N., while in September a mass of them was reported from 48° 30′ to 46° N., between 46° and 50° W. In October a few reports show them within about the same limits as in September, while in November they have almost disappeared, to be followed in December by quite a number from 49° to 46° N., between

46° and 53° W., which were the advanced ice of the present season's prolific output.

From June till November there were thousands of them in and around the Straits of Belle Isle, and the coast of Labrador was full of them. The year 1889 was analyzed at length, for it shows that last year's season has merged into that of the present year without showing any interval. This, however, is explained by the fact that ice did not appear in 1889 in any amount until April, owing possibly to the fact that the season of 1888 was a close one in the Arctic, and the drift of the bergs was very much interrupted and delayed, while in 1889 it was very open and the bergs of the one followed close on the heels of the other. We know that the coast of Labrador was not free of them during the entire year of 1889, so that the drift from the Arctic must have been continued for over a year. The Dundee whalers who spent the summer of 1889 in the Arctic reported a very open season there, with an unusually large number of bergs adrift, and predicted the early appearance and great number of bergs that have been seen since December up to the present time, and it forms an interesting problem to follow out their theory. Unfortunately there are not enough data at hand to go deeply into the subject, but it would appear that an open summer in the Arctic would give an increase in the quantity and size of the bergs during the following spring and summer in low latitudes. In 1888 there were hardly any reports of ice other than in the Straits of Belle Isle and on the coast of Newfoundland, though a few reports show one or two bergs and a little field ice off soundings, but it is hardly worth mentioning, while in 1887 ice was continually present in transatlantic routes from February to August, inclusive.

Of course in different years ice reaches a different southern limit, varying generally between 40° and 41° 30′ N. Bergs will be found between 40° and 55° W., though both of these limits have been exceeded during the present season, while field ice will be found as far as 62° W., owing to the fact that the Gulf of St. Lawrence sends out such large quantities.

The plottings on the accompanying charts were made from reports received at this Office, and the absence of ice above the latitude of 47° or 48° N. is accounted for by the fact that no reports were received from that region, though doubtless it was full of ice during the season.

In the same way many of the bays and coast of Newfoundland were full of ice, as well as the Gulf of St. Lawrence, but few reports were received, owing to the absence of traffic in these parts. The year 1885 is taken as a fair example, though a few bergs may reach a little farther to the eastward than usual, and do not spread quite as far to the westward as the general average.

The deductions drawn as to the drift of ice off soundings have been made from plottings, each mass representing the reports for a month, so by comparing these the above rules were laid down.

It will be seen from the accompanying charts how important it is that all vessels should send reports immediately on their arrival in port of all ice sighted to some central office, where the observations can be platted and recorded in a graphic form, for from these alone can predictions be made. All authentic reports from newspapers, and from other reliable sources should be utilized, and those which would prove invaluable would be from light-house keepers in Newfoundland and Labrador or others on shore, who keep a good record of all ice sighted. Telegraphic reports from northern ports or quick mail facilities will give good results in ample time, but all should go to a central office.

The Hydrographic Office has solicited extensive co-operation on the part of all who would be likely to sight ice, and especially by those who spend the season in ice regions, and the future should bear excellent results.

To this end, a special ice form was prepared and distributed to the Newfoundland sealers, who leave each year on March 10 for the sealing grounds, and who will be the first to sight the heavy Arctic field-ice; to whalers, and the northern fishing fleet. Some of these vessels after the sealing season go to the Arctic whaling grounds and return to port late in the fall, and their reports will be very valuable since they will show the state of the ice in high latitudes.

Arrangements have been made with the light-house keepers of Newfoundland, with many of the Newfoundland and Nova Scotia fishing fleet, and many others, to keep these ice forms and send them to this Office as fast as filled, and in this way an excellent record can be kept of the ice from the time it reaches Cape Chidley until it disappears in low latitudes.

It is unnecessary to dwell upon the importance of the ice problem, or to comment upon the good it will do when more fully understood, as all who encounter it appreciate it already. It stands to reason that the more reports that can be gotten the better will be the predictions. Heretofore almost every report received came from the transatlantic steamers, but in future we hope to secure hundreds of other observers, which, with our old ones, will enable a good showing to be made and produce excellent results.

The Canadian weather service has established a number of stations in and around Hudson Strait, and though only a few years have elapsed since they were started, they are already doing a great deal of good. An important factor in ice observations will be a record of prevailing winds, their force and direction, and current observations. All of these will necessarily be used in making predictions, and temperature of both atmosphere and of the water, on the surface and below, will be very valuable.

The benefit of these will inure to those who make the observations, for thousands of dollars would annually be saved if vessels could know the position of ice before leaving port and steer a course that would keep them clear of it. A rough estimate of the amount saved in coal alone, resulting from delays due to ice, would be over $100,000 per annum, not to mention damage or loss of vessel.

Each month on the North Atlantic Pilot Chart the ice for the previous month is plotted and a course laid down to clear it, which, though some 200 miles longer than the great circle routes from side to side, is the most economical course that can be followed. This takes vessels not only south of the ice region, but an inspection of the charts will show that it generally clears the fog limit, which is specially dangerous in the ice season.

Many vessels have been delayed from a few hours to several days by ice this season from running too far to the northward, while a number have received serious damage, and a few have been lost. A list, which is not complete by any means, is appended showing the number of vessels which have been damaged by ice, and it is safe to say that as many more have been injured of which no report was made public. It will be noticed that few fishing vessels or coasters are mentioned in this list; if they were it would be triple the length it is now.

It is true that many vessels find their interests in the ice-pack, and in the waters where it is most prevalent, but it should be remembered that they are built and strengthened for this work. A wooden ship can venture into ice-fields which would be fatal to one of iron, but if an iron ship be strengthened in her bows, and have a sheathing of wood on the outside, extending several feet above the water-line, she can then venture more boldly into an ice-field. Iron plates in contact with ice are easily cracked, and a weak bow is stove in, showing the imperative necessity of strong water-tight bulk-heads forward. Many a vessel has been saved by these alone. Naturally a short vessel which answers her helm readily will have a better chance of avoiding ice-pans than a longer one that is sluggish after the helm has been put over, and small wooden schooners make safe passage through fields that would cause the loss of the largest and best ocean steamers should they venture on the same track. Much damage and loss of property is annually incurred by the fishing fleet which anchors on the Banks, and from these vessels many valuable reports have been received on the movements of bergs and fields, which are accurate from the fact that the observations were taken from a fixed position.

Frequently a vessel at anchor sights a field coming down on her, say from the northward, which is not heavy enough to necessitate getting up anchor. This field may drift entirely past, and with a southerly wind springing up in the mean time, will drift back to the northward again. Fields have been observed to come from every direction, depending upon current when there is no wind, but following the resultant of wind and current with a moderate breeze.

Fields in drifting past vessels chafe their sides and some cut through, frequently carry away rudders and do other damage. To the trawlers who have their dories out, especially in thick weather, ice-fields cause not only loss of property but loss of life, for it is often impossible to haul the dories back to the vessel, whereupon they are carried away before the ice and the fishermen meet their death from hunger and exposure. Should any doubt exist as to whether a vessel can hold when a field of ice is drifting down on her it is better to slip the cable and run than to take any chances. Small bergs cause damage too by drifting down on vessels at anchor, so a sharp lookout is always necessary during the ice season. Fishermen sometimes take advantage of the presence of bergs to lay in a supply of ice for preserving their fish.

The submarine cables are laid in deep water and not across the Banks, for should a berg ground and come in contact with a cable it would be broken at once.

It has been shown how bergs go to pieces, so it will not be necessary to refer to them again. A high temperature will soften field ice and make it very rotten, so that the slightest motion will cause it to fall to pieces. On reaching the waters of the Gulf Stream, or a warmer atmospheric temperature, it begins to melt, gets soft and spongy, and left in a calm will disappear slowly. But, fortunately, there is seldom a time when there is not a swell on the sea and this soon breaks the pans into small pieces, thus bringing a greater surface in contact with the melting agency. A heavy gale will in a few hours sometimes cause the destruction of a large field by fracture, friction, and continued motion, just as a calm cold night may unite it in a solid mass. Bergs plow their way through fields, break them up, and scatter the pieces, as in the Arctic. Snow preserves them, and often gives the pans the appearance of standing well out of water, and is misleading in this particular. By melting and afterwards freezing it adds to the thickness of the ice.

Before ice is seen from deck the ice blink will often indicate its presence. This is readily understood when it is known that it is caused by the reflection of the rays of light from the sun or moon. On a clear day over the ice on the horizon the sky will be much paler or lighter in color and is easily distinguishable from that overhead, so that a sharp lookout should be kept and changes in the color of the sky noted. Should a vessel not follow the routes laid down by this Office during the ice season and finds herself in the ice, it will often be a saving in time to at once run to the southward or along the trend of the ice-field and round the southern edge, unless a lead can be found through the field. The edges of the field will probably not be found to be uniform in shape, but tongues and capes may extend out in many places, so that it is better to give it a wide berth.

The great danger in attempting to stand through ice lies in the fact that a gale may come up before the ice is cleared, and cause the ice to have such a heavy motion that the bows may be stove in, rudder carried away, or pieces of ice be thrown on deck, or do other damage.

It would seem that underwriters should give better rates to those vessels that keep clear of ice and fog, and the saving in this alone might make up for the expenditure of coal necessary to follow the safer route.

HUGH RODMAN,
Ensign, U. S. Navy.

APPENDIX.

DISASTERS DUE TO ICE.

The following is a partial list of disasters in the North Atlantic due to ice.

1882.

Esquimaux, February 28.—The whaling steamer *Esquimaux* arrived at St. John's, Newfoundland, and reports having been thirteen days among vast fields of ice that stretched southeast and south over 200 miles from the Newfoundland coast. The pack caught her tightly, and she remained in it until she drifted up to St. John's Harbor. At last accounts five steam whalers were visible from Cape Spear locked in the ice and drifting helplessly southward.

Limosa, March 19.—The steam-ship *Limosa* passed through a heavy field of ice and stove in starboard bow.

Rialto, March 29.—The steam-ship *Rialto* ran into a field of ice and stove in both bows; had to steer southeast for 200 miles to clear ice.

Newfoundland, March 27.—The steam-ship *Newfoundland* reports two vessels jammed in the ice 30 miles southwest of Cape Pine.

Promise, March 31.—The schooner *Promise* was struck by a large ice-floe and sunk.

Hermod, April 8.—The steam-ship *Hermod*, in latitude 45° 20' N., longitude 48° W., was surrounded by ice. During a northeast gale and heavy swell ice began to move and stove in bows, filling forward compartments. Bergs all around vessel.

May 10.—Telegrams from St. John's state that there are forty-five sailing vessels and eight steamers caught in the ice in the Gulf of St. Lawrence.

Peruvian, May 19.—The steam-ship *Peruvian*, with 1,000 passengers on board, disabled and locked in the ice at entrance to Gulf of St. Lawrence; in dangerous position; fears are entertained for her safety.

Montreal, May 20.—The steam-ship *Montreal* was hemmed in by ice in Gulf of St. Lawrence for nine days; got clear by passengers and crew cutting an opening.

Harry Wetmore, May 21.—The whaling schooner *Harry Wetmore* reported twenty-one ships locked in ice north of Cape Ray, and two ocean steamers between Capes Ray and Anguille.

Ashdrubal, June 21.—The steam-ship *Ashdrubal* struck a berg 20 miles off Cape Race and sunk.

1883.

Violet, February 19—The steamship *Violet*, when 40 miles east of Louisburg, Cape Breton Island, encountered heavy drift ice. Drifted about in ice for eleven days, during which time a large hole about 6 feet long was knocked in her starboard bow. Arrived at Halifax March 4, forward compartment full of water.

General Birch, March 15—The bark *General Birch*, was found fast in the ice in latitude 45° N., longitude 48° 30' W., with bows stove in, vessel abandoned and full of water.

Christel, March 17—The bark *Christel*, from latitude 47° 20' N., longitude 46° 20' W., to latitude 44° 43' N., longitude 50° 59' W., had considerable metal torn off by ice.

Nettleworth, April 7—The steam-ship *Nettleworth* returned to North Sidney, having had bows stove in by ice off Cape Race Rocks.

Zambesi, May 5—The ship *Zambesi* when 16 miles off Scatari, Cape Breton, struck heavy ice; stove in bows and sank in twenty minutes.

June 7—Report from St. John's, Newfoundland, states that thirty sealing schooners are fast in the ice in the Gulf of St. Lawrence, in northern part.

Barcelona, July 17—The steam-ship *Barcelona*, when 100 miles east of Anticosti, collided with an iceberg, crushing in the whole of her bows 2 feet above water line.

1884.

Notting Hill, January 2—The steam-ship *Notting Hill* collided with an iceberg; was so seriously damaged that she was abandoned January 5, in latitude 46° N., longitude 46° 20' W.

Gloucester, February 24—The steam-ship *Gloucester*, when south of the Banks, had port bow damaged by ice.

George Peabody, February 23—The bark *George Peabody*, in latitude 42° 22' N., longitude 48° 57' W., was abandoned with bows stove in and rudder damaged by ice.

1885.

Ripon City, February 16.—The steam-ship *Ripon City*, when in latitude 45° 30' N., longitude 48° W., in ice-fields, had bow-plates cracked, and had to put into Halifax.

Sussex, February 18.—The steam-ship *Sussex*, in latitude 45° 30' N., longitude 48° W., struck a large cake of ice and stove in bows.

Marance, April 5.—The bark *Marance*, in latitude 46° 30' N., longitude 45° 54' W., was crushed in an ice-floe and sunk.

Young Prince, April 19.—The Newfoundland sealer *Young Prince* collided with an iceberg in Gulf of St. Lawrence and sunk almost immediately.

Moen, May 1.—The bark *Moen* in latitude 46° N., longitude 45° W., collided with a berg and foundered.

Cilurum, May 5.—The steam-ship *Cilurum*, when in latitude 45° N., longitude 47° W., collided with a berg and stove in bows.

Bayard, May 6.—The bark *Bayard*, when in latitude 46° N., longitude 48° W., collided with berg and was abandoned.

Magdalena, May 6.—The bark *Magdalena*, when in latitude 45° N., longitude 47° W., collided with a berg and was abandoned.

Annie Christine, May 7.—The brig *Annie Christine* on Grand Banks struck a berg and foundered.

Mary Louisa, May 10.—The steam-ship *Mary Louisa*, in latitude 49° N., longitude 46° W., fell in with heavy ice, which crushed her plates; foundered.

Jeranos, May 9.—The steam-ship *Jeranos*, from Rotterdam to Montreal, put into North Sydney, Cape Breton, having collided with a berg and knocked a hole in the bow.

City of Berlin, May 19.—The steam-ship *City of Berlin*, in latitude 43° 30' N., longitude 49° 30' W., struck a berg and carried away bowsprit and head-work.

1886.

Lady Agnes, September 20.—The schooner *Lady Agnes*, when 60 miles east of St. John's, Newfoundland, collided with berg, badly damaged; drifted about helplessly for seven days, finally reached St. John's.

1887.

Germania, March 5.—The steam-ship *Germania*, in latitude 48° 09' N., longitude 46° 12' W., from this position southward for four days, ran through heavy ice interspersed with bergs and sustained considerable damage.

Hartville, March 11.—The steam-ship *Hartville,* in latitude 43° 20' N., longitude 47°
W., collided with berg and sustained damage.

Frank A. Williams, March 20.—The schooner *Frank A. Williams* surrounded by ice
20 miles southeast of Sable Island; the pack caused her to leak badly.

Newfoundland, March 24.—The steam-ship *Newfoundland,* when off Cape Race fell
in with heavy ice-field and damaged bow.

1888.

Dove, June 12—The schooner *Dove* was crushed by ice off St. John's.

1889.

Saale, June 11.—The steam-ship *Saale,* in latitude 42° 54' N., longitude 49° 54' W.,
during thick weather collided with berg, but not damaged.

1890.

Nessmore, January 13.—The steam-ship *Nessmore* collided with berg; damaged bows;
narrowly escaped destruction.

Washington City, January 28.—The steam-ship *Washington City* was in ice at intervals between latitude 47° 45' N., longitude 48° 26' W., to latitude 44° 45' N., longitude 51° 42' W.; damaged several bow-plates and filled forward compartment with
water.

Gillett, January 29.—The steam-ship *Gillett,* between latitude 46° 50' N., longitude
46° 45' W., and latitude 45° 32' N., longitude 48° 15' W., collided with ice and knocked
two holes in her bows.

Mareca, February 1 *to* 4.—The steam-ship *Mareca,* between latitude 49° N., longitude
49° 19' W., and latitude 43° 17' N., longitude 50° 36' W., had two plates bent in starboard
bow.

Oliver Emery, February 5.—The bark *Oliver Emery,* in latitude 45° 15' N., longitude
48° 14' W., had been in ice-pack thirty hours, when supplied with provisions by the
steam-ship *Amsterdam.* She had lost a piece of her stern and was leaking badly.

Miranda, February 5.—The steam-ship *Miranda,* from St. John's, Newfoundland, to
Halifax, Nova Scotia, broke the wood lock of her rudder and damaged several bow-
plates while in heavy ice.

Caroline, February 12 *to* 14.—The brig *Caroline,* latitude 43° 55' N., longitude 60°
10' W., had stern badly damaged and sides scraped by field ice.

Meteor, February 17.—The bark *Meteor* spent nine days in an immense ice-field south
of Cape Race. The ice smashed in her bows, carried away the rudder, and opened her
seams. Crew rescued in an exhausted condition by the steam-ship *Marengo* in latitude 43° 7' N., longitude 48° 54' W.

Conscript, February 22.—The steam-ship *Conscript,* from St. John's, Newfoundland, to
Halifax, Nova Scotia, stripped the sheathing from her bows in heavy field ice.

Tynedale, February 22.—The steam-ship *Tynedale,* in latitude 46° 52' N., longitude
47° W., was imprisoned in the ice three days. Started her bow-plates and caused a
bad leak in her collision bulk-head.

Minister Maybach, February 22.—The steam-ship *Minister Maybach,* in latitude 46°
30' N., longitude 46° 44' W., steamed through ice-field eighteen hours; started rivets
in her bow-plates, causing leak in her forward compartment.

Wild Flower, February —.—The steam-ship *Wild Flower,* 300 miles east of Cape Race,
encountered ice, stove in several bow-plates, and filled forward compartment with
water.

India, February —.—The steam-ship *India,* from Baltimore to Hamburg, was damaged
by ice.

Silvia, February —.—The steam-ship *Silvia,* latitude 44° 30' N., longitude 45° 10' W.,
was six days in the ice; had copper and planking badly damaged.

North Cambria, March 1 *to* 3.—The steam-ship *North Cambria*, between latitude 45° 11′ N., longitude 47° 31′ W., and latitude 43° 16′ N., longitude 49° 30′ W., collided with ice during fog; stove in a plate on each bow and filled forward compartment; uarrowly escaped collision with large berg near the last position.

Volunteer, March 10.—The steam-ship *Volunteer*, from St. John's, Newfoundland, for Halifax, Nova Scotia, met heavy ice 40 miles from St. John's; had to steer 125 miles to the southward to clear it; tore off one of her bow-plates.

Lizzie J. Greenleaf, March —.—The schooner *Lizzie J. Greenleaf*, while at anchor on the Grand Bank, had to cut her cable, and lost 100 fathoms of chain, to avoid an iceberg that was drifting down on her.

Kestel, March —.—The schooner *Kestel*, from St. John's, Newfoundland, for Bristol, encountered field ice and damaged sheathing.

Esquimaux, March —.—The sealing steam-ship *Esquimaux*, when northeast of Fogo Island, jammed in heavy ice, damaged bows, and had to return to port.

Strait of Gibraltar, April —.—The steam-ship *Strait of Gibraltar*, from London to New York, put into Louisburg damaged by ice. She was leaking so badly that she had to be grounded to keep her from sinking.

Magdalena, April 16.—The bark *Magdalena*, in latitude 44° 40′ N., longitude 39° W., collided with a berg and was abandoned. Her crew was taken off by the steamship *Umbria*.

Ice in transatlantic routes.

Year.	Field ice. Appeared.	Field ice. Disappeared.	Bergs. Appeared.	Bergs. Disappeared.
1882	February	May	February	August.
1883	....do	April	....do	September.
1884 *	....do	...do	....do	Do.
1885	....do	June	....do	October.
1886	....do	...do	March	August.
1887	....do	May	February	September.
1888 †	....do	June	April	August.
1889	March	August	.. do	November.
1890	January		December, 1889..	

* Newfoundland coast, full till October.
† Very little ice in transatlantic routes; coast of Newfoundland full.

www.ingramcontent.com/pod-product-compliance
Lightning Source LLC
Chambersburg PA
CBHW021459090426
42739CB00009B/1797